This Logbook Belongs to:

_____

Experiment 1

Project Idea Brainstorming

Use these pages to jot down all of your ideas

_____

_____

_____

_____

_____

_____

_____

_____

_____

_____

_____

_____

_____

_____

Project Idea Brainstorming

Use these pages to jot down all of your ideas

_____

_____

_____

_____

_____

_____

_____

_____

_____

_____

_____

_____

_____

Project Idea Brainstorming

Use these pages to jot down all of your ideas

_____

_____

_____

_____

_____

_____

_____

_____

_____

_____

_____

_____

_____

_____

_____

Can I Make This Project Work?

Use the following questions to determine if your project idea is a good one for the science fair. The answer to all of the questions should be yes.

	Yes	No
Can I write a question for my topic?	[ ]	[ ]
Can it be tested?	[ ]	[ ]
Do I have enough time to test it?	[ ]	[ ]
Can I get all of the materials I need?	[ ]	[ ]
Can I afford the materials?	[ ]	[ ]
Is it safe?	[ ]	[ ]
Is it original?	[ ]	[ ]
Do I fully understand the project?	[ ]	[ ]
Can it be researched?	[ ]	[ ]
Does it interest me?	[ ]	[ ]

Thoughts

Thoughts

Narrow It Down

Three testable questions for my project

1._____

_____

_____

2. _____

_____

_____

3. _____

_____

_____

Resources Log

Books/Magazines/Newspapers

Title	Author	Date

Resources Log

Websites

URL	Author	Date

Resources Log

Personal Interviews

Name	Contact Info	Date

Statement of Purpose

What is the purpose of your experiment? What are you looking to find out?

_____

_____

_____

_____

_____

_____

_____

_____

_____

_____

_____

_____

_____

Hypothesis

What do you think will be the result of your experiment?

_____

_____

_____

_____

_____

_____

_____

_____

_____

_____

_____

_____

_____

_____

Materials

What will you use to complete your experiment?

_____

_____

_____

_____

_____

_____

_____

_____

_____

_____

_____

_____

_____

_____

Procedure

What will you do to perform the experiment?

1._____

_____

_____

2. _____

_____

_____

3. _____

_____

_____

4. _____

_____

_____

5. _____

_____

_____

Procedure

What will you do to perform the experiment?

6. _____

_____

_____

7. _____

_____

_____

8. _____

_____

_____

9. _____

_____

_____

10. _____

_____

_____

Procedure

What will you do to perform the experiment?

11. _____

_____

_____

12. _____

_____

_____

13. _____

_____

_____

14. _____

_____

_____

15. _____

_____

_____

Procedure

What will you do to perform the experiment?

16. _____

_____

_____

17. _____

_____

_____

18. _____

_____

_____

19. _____

_____

_____

20. _____

_____

_____

Observations and Results

What did you see as you completed your experiment?

Observations and Results

What did you see as you completed your experiment?

Observations and Results

What did you see as you completed your experiment?

Observations and Results

What did you see as you completed your experiment?

_____

_____

_____

_____

_____

_____

_____

_____

_____

_____

_____

_____

_____

_____

Observations and Results

What did you see as you completed your experiment?

_____

_____

_____

_____

_____

_____

_____

_____

_____

_____

_____

_____

_____

_____

Observations and Results

What did you see as you completed your experiment?

_____

_____

_____

_____

_____

_____

_____

_____

_____

_____

_____

_____

_____

_____

Observations and Results

What did you see as you completed your experiment?

_____

_____

_____

_____

_____

_____

_____

_____

_____

_____

_____

_____

_____

_____

Observations and Results

What did you see as you completed your experiment?

_____

_____

_____

_____

_____

_____

_____

_____

_____

_____

_____

_____

_____

_____

Observations and Results

What did you see as you completed your experiment?

_____

_____

_____

_____

_____

_____

_____

_____

_____

_____

_____

_____

_____

_____

Conclusion

What did you discover as a result of your experiment? Was your hypothesis correct? Why or why not?

_____

_____

_____

_____

_____

_____

_____

_____

_____

_____

_____

_____

_____

_____

Conclusion

What did you discover as a result of your experiment? Was your hypothesis correct? Why or why not?

_____

_____

_____

_____

_____

_____

_____

_____

_____

_____

_____

_____

_____

_____

Conclusion

What did you discover as a result of your experiment? Was your hypothesis correct? Why or why not?

_____

_____

_____

_____

_____

_____

_____

_____

_____

_____

_____

_____

_____

_____

Conclusion

What did you discover as a result of your experiment? Was your hypothesis correct? Why or why not?

_____

_____

_____

_____

_____

_____

_____

_____

_____

_____

_____

_____

_____

_____

_____

Experiment 2

Project Idea Brainstorming

Use these pages to jot down all of your ideas

_____

_____

_____

_____

_____

_____

_____

_____

_____

_____

_____

_____

Project Idea Brainstorming

Use these pages to jot down all of your ideas

_____

_____

_____

_____

_____

_____

_____

_____

_____

_____

_____

_____

Project Idea Brainstorming

Use these pages to jot down all of your ideas

_____

_____

_____

_____

_____

_____

_____

_____

_____

_____

_____

_____

_____

Can I Make This Project Work?

Use the following questions to determine if your project idea is a good one for the science fair. The answer to all of the questions should be yes.

	Yes	No
Can I write a question for my topic?	[ ]	[ ]
Can it be tested?	[ ]	[ ]
Do I have enough time to test it?	[ ]	[ ]
Can I get all of the materials I need?	[ ]	[ ]
Can I afford the materials?	[ ]	[ ]
Is it safe?	[ ]	[ ]
Is it original?	[ ]	[ ]
Do I fully understand the project?	[ ]	[ ]
Can it be researched?	[ ]	[ ]
Does it interest me?	[ ]	[ ]

Thoughts

Thoughts

_____

_____

_____

_____

_____

_____

_____

_____

_____

_____

_____

_____

_____

_____

_____

Narrow It Down

Three testable questions for my project

1. _____

_____

_____

2. _____

_____

_____

3. _____

_____

_____

Resources Log

Books/Magazines/Newspapers

Name	Contact Info	Date

Resources Log

Websites

Name	Contact Info	Date

Resources Log

Personal Interviews

Name	Contact Info	Date

Statement of Purpose

What is the purpose of your experiment? What are you looking to find out?

_____

_____

_____

_____

_____

_____

_____

_____

_____

_____

_____

_____

_____

_____

Hypothesis

What do you think will be the result of your experiment?

_____

_____

_____

_____

_____

_____

_____

_____

_____

_____

_____

_____

_____

_____

Materials

What will you use to complete your experiment?

_____

_____

_____

_____

_____

_____

_____

_____

_____

_____

_____

_____

_____

Procedure

What will you do to perform the experiment?

1. _____

_____

_____

2. _____

_____

_____

3. _____

_____

_____

4. _____

_____

_____

5. _____

_____

_____

Procedure

What will you do to perform the experiment?

6. _____

_____

_____

7. _____

_____

_____

8. _____

_____

_____

9. _____

_____

_____

10. _____

_____

_____

Procedure

What will you do to perform the experiment?

11. _____

_____

_____

12. _____

_____

_____

13. _____

_____

_____

14. _____

_____

_____

15. _____

_____

_____

Procedure

What will you do to perform the experiment?

16. _____

_____

_____

17. _____

_____

_____

18. _____

_____

_____

19. _____

_____

_____

20. _____

_____

_____

Observations and Results

What did you see as you completed your experiment?

Observations and Results

What did you see as you completed your experiment?

Observations and Results

What did you see as you completed your experiment?

Observations and Results

What did you see as you completed your experiment?

_____

_____

_____

_____

_____

_____

_____

_____

_____

_____

_____

_____

_____

_____

Observations and Results

What did you see as you completed your experiment?

_____

_____

_____

_____

_____

_____

_____

_____

_____

_____

_____

_____

_____

_____

Observations and Results

What did you see as you completed your experiment?

_____

_____

_____

_____

_____

_____

_____

_____

_____

_____

_____

_____

_____

_____

_____

Observations and Results

What did you see as you completed your experiment?

_____

_____

_____

_____

_____

_____

_____

_____

_____

_____

_____

_____

_____

_____

Observations and Results

What did you see as you completed your experiment?

_____

_____

_____

_____

_____

_____

_____

_____

_____

_____

_____

_____

_____

_____

Observations and Results

What did you see as you completed your experiment?

_____

_____

_____

_____

_____

_____

_____

_____

_____

_____

_____

_____

_____

_____

Conclusion

What did you discover as a result of your experiment? Was your hypothesis correct? Why or why not?

_____

_____

_____

_____

_____

_____

_____

_____

_____

_____

_____

_____

_____

_____

Conclusion

What did you discover as a result of your experiment? Was your hypothesis correct? Why or why not?

_____

_____

_____

_____

_____

_____

_____

_____

_____

_____

_____

_____

_____

_____

_____

Conclusion

What did you discover as a result of your experiment? Was your hypothesis correct? Why or why not?

_____

_____

_____

_____

_____

_____

_____

_____

_____

_____

_____

_____

_____

Conclusion

What did you discover as a result of your experiment? Was your hypothesis correct? Why or why not?

_____

_____

_____

_____

_____

_____

_____

_____

_____

_____

_____

_____

_____

_____

Experiment 3

Project Idea Brainstorming

Use these pages to jot down all of your ideas

_____

_____

_____

_____

_____

_____

_____

_____

_____

_____

_____

_____

_____

_____

Project Idea Brainstorming

Use these pages to jot down all of your ideas

_____

_____

_____

_____

_____

_____

_____

_____

_____

_____

_____

_____

_____

_____

Project Idea Brainstorming

Use these pages to jot down all of your ideas

_____

_____

_____

_____

_____

_____

_____

_____

_____

_____

_____

_____

_____

_____

Can I Make This Project Work?

Use the following questions to determine if your project idea is a good one for the science fair. The answer to all of the questions should be yes.

	Yes	No
Can I write a question for my topic?	[ ]	[ ]
Can it be tested?	[ ]	[ ]
Do I have enough time to test it?	[ ]	[ ]
Can I get all of the materials I need?	[ ]	[ ]
Can I afford the materials?	[ ]	[ ]
Is it safe?	[ ]	[ ]
Is it original?	[ ]	[ ]
Do I fully understand the project?	[ ]	[ ]
Can it be researched?	[ ]	[ ]
Does it interest me?	[ ]	[ ]

Thoughts

Thoughts

Narrow It Down

Three testable questions for my project

1._____

_____

_____

2. _____

_____

_____

3. _____

_____

_____

Resources Log

Books/Magazines/Newspapers

Name	Contact Info	Date

Resources Log

Websites

Name	Contact Info	Date

Resources Log

Personal Interviews

Name	Contact Info	Date

Statement of Purpose

What is the purpose of your experiment? What are you
looking to find out?

_____

_____

_____

_____

_____

_____

_____

_____

_____

_____

_____

_____

Hypothesis

What do you think will be the result of your experiment?

_____

_____

_____

_____

_____

_____

_____

_____

_____

_____

_____

_____

_____

Materials

What will you use to complete your experiment?

_____

_____

_____

_____

_____

_____

_____

_____

_____

_____

_____

_____

_____

_____

Procedure

What will you do to perform the experiment?

1. _____

_____

_____

2. _____

_____

_____

3. _____

_____

_____

4. _____

_____

_____

5. _____

_____

_____

Procedure

What will you do to perform the experiment?

6. _____

_____

_____

7. _____

_____

_____

8. _____

_____

_____

9. _____

_____

_____

10. _____

_____

_____

Procedure

What will you do to perform the experiment?

11. _____

_____

_____

12. _____

_____

_____

13. _____

_____

_____

14. _____

_____

_____

15. _____

_____

_____

Procedure

What will you do to perform the experiment?

16. _____

_____

_____

17. _____

_____

_____

18. _____

_____

_____

19. _____

_____

_____

20. _____

_____

_____

Observations and Results

What did you see as you completed your experiment?

Observations and Results

What did you see as you completed your experiment?

Observations and Results

What did you see as you completed your experiment?

Observations and Results

What did you see as you completed your experiment?

_____

_____

_____

_____

_____

_____

_____

_____

_____

_____

_____

_____

_____

_____

Observations and Results

What did you see as you completed your experiment?

_____

_____

_____

_____

_____

_____

_____

_____

_____

_____

_____

_____

_____

Observations and Results

What did you see as you completed your experiment?

_____

_____

_____

_____

_____

_____

_____

_____

_____

_____

_____

_____

_____

_____

Observations and Results

What did you see as you completed your experiment?

_____

_____

_____

_____

_____

_____

_____

_____

_____

_____

_____

_____

_____

_____

Observations and Results

What did you see as you completed your experiment?

_____

_____

_____

_____

_____

_____

_____

_____

_____

_____

_____

_____

_____

_____

Observations and Results

What did you see as you completed your experiment?

_____

_____

_____

_____

_____

_____

_____

_____

_____

_____

_____

_____

_____

Conclusion

What did you discover as a result of your experiment? Was your hypothesis correct? Why or why not?

_____

_____

_____

_____

_____

_____

_____

_____

_____

_____

_____

_____

_____

_____

Conclusion

What did you discover as a result of your experiment? Was your hypothesis correct? Why or why not?

_____

_____

_____

_____

_____

_____

_____

_____

_____

_____

_____

_____

_____

_____

Conclusion

What did you discover as a result of your experiment? Was your hypothesis correct? Why or why not?

_____

_____

_____

_____

_____

_____

_____

_____

_____

_____

_____

_____

_____

_____

Conclusion

What did you discover as a result of your experiment? Was your hypothesis correct? Why or why not?

_____

_____

_____

_____

_____

_____

_____

_____

_____

_____

_____

_____

_____

Conclusion

What did you discover as a result of your experiment? Was
your hypothesis correct? Why or why not?

_____

_____

_____

_____

_____

_____

_____

_____

_____

_____

_____

_____

_____

Conclusion

What did you discover as a result of your experiment? Was your hypothesis correct? Why or why not?

_____

_____

_____

_____

_____

_____

_____

_____

_____

_____

_____

_____

_____

_____

www.ingramcontent.com/pod-product-compliance
Lightning Source LLC
Chambersburg PA
CBHW080231180526
45158CB00010BA/3013